MURDER OF MULTITUDES

PROLOGUE - COVID AND CLIMATE CHRONICLES - THE BIG CULL

*Imagine that you understood everything from the start, and warned
everyone from March 2020 about the Covid-19 plandemic / scamdemic:
NO harmful Lockdowns!
and
NO toxic Vaccines!*

*Nobody listened, trillions of dollars were wasted, tens of millions of people
were killed by improper medical treatment and the toxic Covid-19 "vaccines",
billions more were vax-injured, and the carnage is far from over...*

*That was my objective, my full-time mission since February 2020:
- to tell the truth, based on the facts; to save lives, millions of lives.
I published early, accurate information to try to protect everyone.
Nobody listened. I might have saved a handful of people.*

- Allan MacRae, Calgary

MURDER OF MULTITUDES

INTRODUCTION - COVID AND CLIMATE CHRONICLES - THE BIG CULL

I identified the Covid-19 Lockdowns and Vaccines frauds starting in Feb2020 and published on 21Mar2020. We called the Climate and Green Energy scams starting in 2002. All correct!

This is my book about the Covid and Climate frauds. This is why I wrote it:

Fine young faces, our future, killed by the Covid-19 vaxes. I warned you in early 2020!

I've watched this carnage unfold - identified it in Feb2020, published on 21Mar2020, and was ignored. 13 million Covid-19 vax-caused deaths in the Western world to end 2022, 19 million by end 2023, 40 million worldwide – and it's far from over! The vax-deaths will continue for years.

Save lives! Safeguard everyone you love. Stop all toxic Covid-19 vaxes. Treat the vaxed now.

Best regards to all, Allan MacRae, Calgary

THE CONTENTS OF THE BOOK PRIMARILY COMPRISE:

- Letters to governments and media informing them since 21Mar2020 about the Covid-19 Lockdown and Vaccine scams.
- Papers published from Nov2002 to present on the Climate and Green Energy scams.
- Comments on The Corruption of our Institutions, beyond the Covid and Climate scams.

COVID & CLIMATE CHRONICLES - THE BIG CULL

"The ability to correctly predict is the best objective measure of scientific and technical competence."
Our predictions on the Covid and Climate scams are among the earliest and most accurate.

Our scientific predictions and conclusions on both Covid and Climate are infinitely more accurate than the mainstream narratives, which are false and baselessly alarmist.

The Chapters in this book reflect the current data at the time those Chapters were written.

All the pages of this book are copyright, the property of Allan M R MacRae, Calgary. Exceptions are all articles written by others.

EDITORIAL

These facts in this book prove why our incompetent politicians should have listened to my early warnings about the Covid and Climate frauds. Objectively, my career accomplishments and the resulting benefits to Canada exceed that of any living Canadian politician.
Our most respected Alberta and Canadian politicians owe their career success to the economic boom that I and a few colleagues created. Politicians for the past two decades received credit for that economic success but contributed nothing to it, and often sabotaged it with their utter incompetence and destructive policies on Climate, Green Energy, harmful Covid-19 Lockdowns, toxic Covid-19 "Vaccines", and runaway government spending and money-printing.

We are governed by scoundrels and imbeciles, and our once-great country is failing.

MURDER OF MULTITUDES

CONTENTS

ABOUT THE AUTHOR

Allan M.R. MacRae, B.A.Sc.(Eng.) Queen's U, M.Eng. U of Alberta

https://energy-experts-international.com

Conducted business at a senior level on six of the world's seven continents.

Recommended AGAINST the Covid-19 lockdowns on 21March2020, in a post essentially identical to the Great Barrington Declaration published 6 months later by the world's top experts.

Highlighted the significant risks of the experimental Pfizer and Moderna Covid-19 mRNA injections and the high risk-low reward of these injections to under-65's and especially to very-low-risk groups including schoolchildren, in a letter to politicians and media on 8January2021.

Safeguarded up to 600,000 Calgarians from probable death by intervening to shut down the Mazeppa sour gas project; MacRae was honored by the Society of Petroleum Engineers.

Authored ~12 papers since 2002 that prove catastrophic human-made global warming is a false crisis, and intermittent green energy is not a practical solution.

Advocated since 2002 against the fraudulent claims of leftist extremists that have cost Alberta and Canada over $120 billion in lost oil revenues.

Wrote the competent Energy Policy for the Wild Rose Party, rejecting the destructive Stelmach Royalty changes.

Initiated the New Oilsands Royalty Terms and the New Oilsands Tax Terms implemented by the Klein PC's and the Feds and also the major reduction of Syncrude Canada Ltd. production costs.

Incorporated these initiatives into a comprehensive strategy for Syncrude Canada, which was implemented and was instrumental in the successful growth of the Alberta oil sands. **Canada became the 4th largest oil producer in the world, with $250 billion ($500+ billion today) capital investment in Alberta and 500,000 new jobs created in Canada. Canada became the largest foreign supplier of energy to the USA and the strongest economy in the G8.**

SUMMARY - COVID & CLIMATE CHRONICLES - THE BIG CULL

"The ability to correctly predict is the best objective measure of scientific and technical competence."
Our predictions and conclusions on the Covid and Climate scams are among the earliest and most accurate. I'm just trying to save lives – hundreds of millions of lives!

This is my book about the Covid and Climate frauds. This is why I wrote it:
Fine young faces, our future, killed by the Covid-19 vaxes. I warned you in early 2020!

I've watched this carnage unfold - identified it in Feb2020, published on 21Mar2020, and was ignored. 13 million Covid-19 vax-deaths in the Western world to end 2022, 19 million by end 2023, 40 million worldwide – and it's far from over! The vax-deaths will continue for years.

Safeguard everyone you love. No more toxic Covid-19 vaxes. Treat the vaxed now.

Best regards to all, Allan MacRae in Calgary

THE CONTENTS OF THIS BOOK PRIMARILY COMPRISE:

- My public warnings since 21Mar2020 on the Covid-19 Lockdown and Vaccine scams.
- Papers published from Nov2002 to present on the Climate and Green Energy scams.
- Observations on The Corruption of our Institutions.

See also my earlier publications that include the proofs that I removed from this short book.

SCIENTIFIC COMPETENCE - THE ABILITY TO CORRECTLY PREDICT

by Allan MacRae October 20, 2021 to Present

COVID & CLIMATE CHRONICLES - THE BIG CULL

I called the Covid-19 Lockdowns and Vaccines scams in Feb2020 and published on 21Mar2020. We called the Climate and Green Energy scams in 2002. All correct!

The ability to correctly predict is the best objective measure of scientific and technical competence.

Our scientific predictions and conclusions on both these subjects are infinitely more accurate than the mainstream narratives, which are false and baselessly alarmist.

COVID & CLIMATE CHRONICLES - THE BIG CULL

The Climate and Covid scams were "Crimes Against Humanity" – wars against scientific reason and technical competence that exceed the 80 million souls killed on all sides in World War 2.

THE CLIMATE CULL

My co-authors and I wrote the following correct observations about the Global Warming (aka "Climate" aka "CAGW") and Green Energy scams in 2002:

1. "Climate science does not support the theory of catastrophic human-made global warming – the alleged warming crisis does not exist."

2. "The ultimate agenda of pro-Kyoto advocates is to eliminate fossil fuels, but this would result in a catastrophic shortfall in global energy supply – the wasteful, inefficient energy solutions proposed by Kyoto advocates simply cannot replace fossil fuels."

– by Sallie Baliunas (Astrophysicist, Harvard-Smithsonian), Tim Patterson (Paleoclimatologist, Carleton U), Allan MacRae (Professional Engineer, retired, Queen's U, U of Alberta)

Nothing has changed in the intervening 20+ years since we published our article, except the huge costs of this global-scale fraud:

- tens of trillions of dollars of scarce global resources have been squandered on "wasteful, inefficient" green energy scams;
- hundreds of millions of lives have been wasted, especially in the "developing world", by denying them access to fossil fuel energy;
- most of our leaders, who are scientifically uneducated, have adopted the "Global Warming" and "Net Zero" falsehoods;
- the same Climate fraudsters are now attacking our food supplies, again to allegedly fight fictitious Global Warming.

There never was any scientific or technical support for the Global Warming and Green Energy scams - it's always been a false propaganda campaign concocted by extremists for their own financial and political gain.

My TV interview of 15May2023 with Laura-Lynn Tyler Thompson: Allan MacRae (lauralynn.tv) on the Covid and Climate scams. My interview runs from 22:00 to 54:00.

My TV interview of 12Apr2023 on the Climate scam:
https://rumble.com/v2jhsxu-talk-truth-allan-macrae.html

My TV interview of 14Apr2023 with Dr Roger Hodkinson on the Covid and Climate scams:
https://rumble.com/v2i4rmm-c19v-death-stats-dr.-roger-hodkinson-and-allan-mcrae-tpc-1181.html

THE COVID CULL

In 2020, humanity was again violently assaulted, this time by the Covid-19 Lockdowns and "Vaccines" scam. I correctly called this scam in Feb2020, within the first month of its public existence, and published on 21Mar2020:

21March2020
"LET'S CONSIDER AN ALTERNATIVE APPROACH:

~~Isolate~~ Protect people over sixty-five and those with poor immune systems and return to business-as-usual for people under sixty-five.

This will allow "herd immunity" to develop much sooner and older people will thus be more protected AND THE ECONOMY WON'T CRASH."

Six months later world experts published the same recommendations in the Great Barrington Declaration.

On 8Jan2021 I wrote government representatives and media and strongly advised:

"The Covid-19 vaccine developments were rushed and are not proven safe or effective and should NOT be taken, especially by the low-risk population - those under-65 or recovered from Covid-19. The two experimental Covid-19 vaccines that contain mRNA (Pfizer and Moderna) are especially risky – due to unknown future side-effects, the risk-to-reward is far too high for the low-risk group."

Two years later the Surgeon General of Florida made the same recommendation.

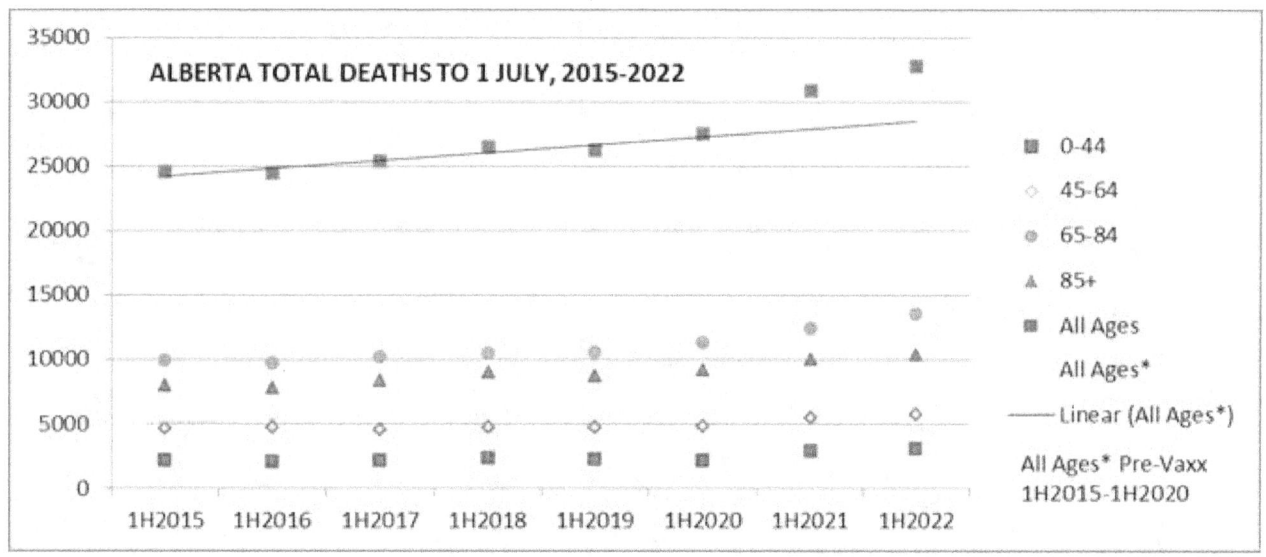

I first published this graph in 2021 – **IMPORTANT: there was no major increase in Total Deaths in Alberta to 1Jul2020, the end of the first Covid-19 flu season. That means NO real pandemic!**

Total Deaths are typically much greater in the Winter flu season every year.

The reason there was NO significant total death increase in Alberta in the 12 months from 1Jul2019 to 30Jun2020, the first "Covid-19 flu year", was because competent Alberta physicians practiced early treatment. Pre-Covid 2017-2018 was a worse flu-year than Covid flu-year 2019-2020 for total deaths. **This proves that the Covid-19 illness was not a dangerous plague, was not fatal to the general population and the panicked over-reaction to Covid-19 was wrong, and needlessly cost trillions of dollars and millions of lives.**

NO EXCESS DEATHS MEANS NO REAL DANGEROUS COVID-19 PANDEMIC!

The Covid-19 "vaccines" were deployed in Alberta in Jan2021. There was a large increase in total deaths for all ages AFTER the toxic Covid-19 injections were deployed.

The big increase in Total Deaths happened by 1H2021 and was caused by the toxic Covid-19 "vaccines". There is credible evidence that this carnage was not an error, but was known to insiders in advance.

Based on total Alberta total deaths:
Deaths attributed to the toxic Covid-19 "vaccines" in 2021 & 2022 totaled 98,000 Canadians to end 2022, increasing to 145,000 by end2023. Justin, Jagmeet and their corrupt leftist cohorts have now killed more Canadians than the 105,000 lost in WW1 & WW2.

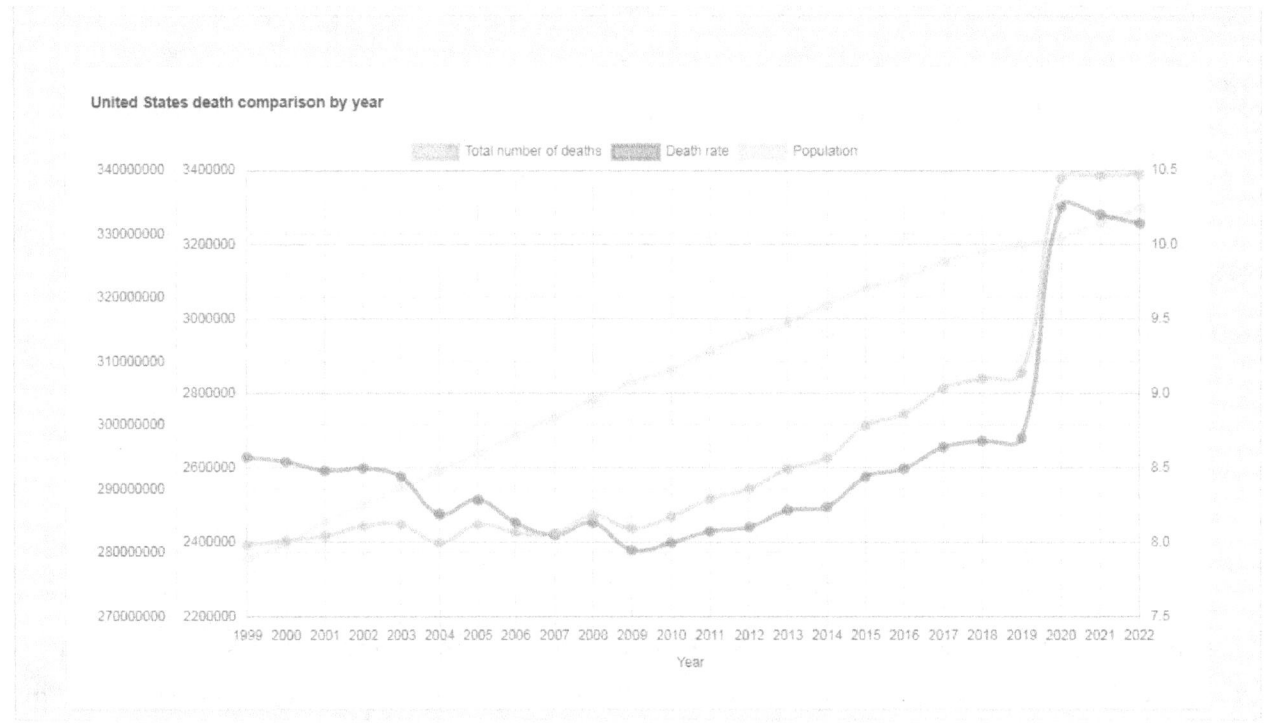

Based on USA total deaths:
Incompetent, late Covid-19 treatment, Remdesivir and ventilators caused ~530,000 mostly-preventable American deaths in 2020. The toxic Covid-19 "vaccines" caused an additional ~1.1 million American deaths in 2021 and 2022, equal to all the deaths in all of America's wars back to 1776. The death toll in the USA from the mismanagement of the Covid-19 illness and the toxic Covid-19 "vaccines" totaled ~1.6 million to end 2022, and it is far from over.

Based on total England and Wales deaths, I ran an approximate analysis and **calculated total British Covid-19 vax-caused deaths in 2021 and 2022 of 172,000.** I posted that message at Metro.UK and it was censored after one day.

On 9Feb2023, based on Alberta total deaths, I calculated 12.9 million deaths worldwide caused by the toxic Covid-19 "vaccines" and this total was independently confirmed by Rancourt et al based on Australian and Israeli total death data. I estimated that total vax-caused deaths would increase to 19 million by end2023. These totals exclude Covid-19 vax-caused-deaths in China, India and Russia.

In the Western world, the Covid-19 vax-caused death toll increases by ~500,000 per month. Globally, including China, India and Russia, total vax-caused-deaths increase by approx. one million per month. I estimate approx. 40 million vax-caused-deaths globally by the end of 2023... and so it goes. These Covid-19 vax-caused-deaths will continue for many years.

IMPORTANT: Video of Denis Rancourt's interview by the National Citizens' Covid-19 Inquiry. Denis Rancourt expert testimony National Citizens Inquiry - Ottawa 17 May 2023 (odysee.com). It is excellent! Rancourt also states: **NO REAL PANDEMIC!**

Since Nov2022 I have advocated simple, over-the-counter worldwide treatments of the vaxed that would save millions of lives.

My initial posts on this book summarize the important points of these global scams. Subsequent chapters include my many letters to politicians and media who were complicit in these crimes against humanity. Nuremberg 2.0!

Best regards to all, Allan MacRae, Calgary

Verily I say unto you, Inasmuch as ye have done it unto one of the least of these my brethren, ye have done it unto me.
- Matthew 25:40

This is my creed. It was the creed of my father and my grandfathers - they lived it.

THE CLIMATE AND GREEN ENERGY SCAMS
Selected Political and Scientific Papers - Nov2002 to Present

The Climate and Covid scams were Crimes Against Humanity – wars against scientific reason and technical competence that exceed the 80 million souls killed on all sides in World War 2.

My co-authors and I wrote the following correct observations about the Global Warming (aka "Climate" aka "CAGW") and Green Energy scams in 2002:

1.   ***"Climate science does not support the theory of catastrophic human-made global warming – the alleged warming crisis does not exist."***

2.   ***"The ultimate agenda of pro-Kyoto advocates is to eliminate fossil fuels, but this would result in a catastrophic shortfall in global energy supply – the wasteful, inefficient energy solutions proposed by Kyoto advocates simply cannot replace fossil fuels."***
– by Sallie Baliunas (Astrophysicist, Harvard-Smithsonian), Tim Patterson (Paleoclimatologist, Carleton U), Allan MacRae (Professional Engineer, retired, Queen's U, U of Alberta)

Nothing has changed in the intervening 20+ years since we published our article, except the huge costs of this global-scale fraud:

- tens of trillions of dollars of scarce global resources have been squandered on "wasteful, inefficient" green energy scams;
- hundreds of millions of lives have been wasted, especially in the "developing world", by denying them access to fossil fuel energy;
- most of our leaders, who are scientifically uneducated, have adopted the "Global Warming" and "Net Zero" falsehoods;
- the same Climate fraudsters are now attacking our food supplies, again to allegedly fight fictitious Global Warming.

There never was any scientific or technical support for the Global Warming and Green Energy scams - it's always been a false propaganda campaign concocted by extremists to harm our economies and promote their totalitarian agenda.

Here is my TV interview of 12Apr2023 on the Climate scam:

https://rumble.com/v2jhsxu-talk-truth-allan-macrae.html

Here is my TV interview of 14Apr2023 with Dr Roger Hodkinson on the Covid and Climate scams:

https://rumble.com/v2i4rmm-c19v-death-stats-dr.-roger-hodkinson-and-allan-mcrae-tpc-1181.html

Here is my TV interview of 15May2023 (runs from 22:00 to 54:00) on the Covid and Climate scams:

Laura-Lynn Tyler Thompson: Allan MacRae (lauralynn.tv)

Selected Political and Scientific Papers - Nov2002 to Present
(Open link for more papers)

THE GREAT RESET: PLANNING THE THEFT OF CANADA?
by Robert McCarter October 2, 2022
https://fcpp.org/2022/10/02/the-great-reset-planning-the-theft-of-canada/
"He explained, expecting me to agree, that Canada's present economic model was seriously flawed and had to be replaced. I bit my tongue. He continued, people expected too much, unregulated consumerism was unsustainable and Canadians would have to learn to make do with less. The government would have to take more control over people's lives and enforce an austere lifestyle. The present high economic expectations are the enemy and we would have to have a strong global government that would redistribute the wealth to poorer nations. Fossil fuels would be phased out on an accelerated timetable and air travel would be limited to need."

NO EVIDENCE OF CLIMATE CRISIS
by Allan M.R. MacRae, B.A.Sc., M.Eng., June 23, 2022
https://www.worldcommercereview.com/html/macrae-no-evidence-of-a-climate-crisis.html
"To conclude, the alleged fossil-fuel-caused Global Warming Crisis does not exist in reality. The only real, measurable impact of increasing atmospheric CO_2 concentrations is improved crop yields – which are hugely beneficial."

THE REAL COST OF LOCKDOWNS

I independently published the correct call on the Covid-19 Lockdowns on 21Mar2020. one of the earliest and most accurate assessments anywhere.

I identified the Covid-19 scam in Feb2020 - Covid-19 was only fatal to the very elderly and infirm - but I was reluctant to publish until I learned on ~20Mar2020 that all our hospitals were EMPTY! That's when I called the Covid scam and published:

21Mar2020

"LET'S CONSIDER AN ALTERNATIVE APPROACH:

Isolate people over sixty-five and those with poor immune systems and return to business-as-usual for people under sixty-five.

This will allow "herd immunity" to develop much sooner and older people will thus be more protected AND THE ECONOMY WON'T CRASH."

22Mar2020

"This full-lockdown scenario is especially hurting service sector businesses and their minimum-wage employees - young people are telling me they are "financially under the bus". The young are being destroyed to protect us over-65's. A far better solution is to get them back to work and let us oldies keep our distance, and get "herd immunity" established ASAP - in months not years. Then we will all be safe again."

Six months later on 4Oct2020, world experts published the same recommendation in their Great Barrington Declaration.

As rare as it was, this correct very-early call was not that difficult – most of the early data was junk, but there was credible mortality data from the Diamond Princess cruise ship and a few other good sources.

I cannot believe that the professional epidemiologists were all that incompetent - the full-Gulag lockdown looked more and more like a huge global fraud - a hugely-funded global fear propaganda campaign inconsistent with the facts, like the Global Warming/Green Energy scam – and for the same totalitarian political objectives.

The full lockdown was originally intended to prevent the "tsunami of cases from swamping out medical system" – A TSUNAMI OF CASES that NEVER HAPPENED! Medical people knew this reality by about late-March2020, ~two weeks into the lockdown, but our Alberta hospitals were emptied for over two months!

Since then, the lockdown has been extended through two years, and has squandered trillions of dollars and harmed billions of people, and for what? The lockdown has NOT saved lives – all it has done is prolong the life of the virus by delaying herd immunity - which allowed the virus to survive into the next flu season. It also allowed the virus to mutate to more contagious forms.

Forget vaccines. Every flu virus in human history dies the same way – via herd immunity – and the lockdowns and vaccines only served to delay herd immunity, drive Covid-19 variants (Darwin 1.0) and cause enormous economic and human harm.

Sweden did it mostly right (NO lockdown) and has achieved herd immunity, and most other countries did it entirely wrong.

It was obvious by ~1March2020 that Covid-19 was a relatively mild flu, considerably milder than the 2017-2018 seasonal flu and others, and essentially only dangerous to the very elderly and very infirm. The workforce and students were barely affected – many were asymptomatic.

It is really difficult to believe that supposedly intelligent "medical experts" could be this stupid for this long – the full-Gulag lockdown looked more like a deliberate panic-driven scam every passing day.

"*At a high level, corruption stems from the systemic disrespect of a nation's institutions and legal system, especially by those in power. . . .The hallmark of virtually all corruption is 'playing dumb'.*"

HUGE AVOIDABLE LATE-TREATMENT AND VACCINE-CAUSED COVID DEATHS - WORLD, USA, CANADA – posted at DeMed Medical Website

I estimated 13 million deaths worldwide were caused by the toxic Covid-19 "vaccines" to end 2022, increasing to 19 million by end 2023. The 13 million total was confirmed by Rancourt et al (Feb2023).

I posted the following on 16Mar2023 on this medical website:

https://forum.demed.com/COVID/posts/oYIBb1xKGgR2JtImpbTy

and emailed it to Alberta and North American politicians and media

and posted it on Twitter:

The death toll in the USA from the mismanagement of the Covid-19 illness and the toxic Covid-19 "vaccines" totaled 1.61 million to end 2022. The Covid-19 avoidable American death toll to date exceeds the total 1.1 million American deaths in all their wars.

Deaths attributed to the toxic Covid-19 "vaccines" in 2021&2022 totaled 98,000 Canadians to end 2022, increasing to 145,000 by end 2023. Justin, Jagmeet and their Liberal and NDP cohorts have now killed more Canadians than the 105,000 we lost in WW1 and WW2.

On 8Jan2021 I wrote to Alberta Premier Kenney and every Alberta MLA and many federal and civic politicians the following strong recommendation, which has also proved correct – **NO COVID-19 VACCINES!**:

"The Covid-19 vaccine developments were rushed and are not proven safe or effective and should NOT be taken, especially by the low-risk population - those under-65 or recovered from Covid-19. The two experimental Covid-19 vaccines that contain mRNA (Pfizer and Moderna) are especially risky – due to unknown future side-effects, the risk-to-reward is far too high for the low-risk group."

Two years later the Surgeon General of Florida published the same conclusion.

Robert W Malone Feb 17 2023

[excerpt]

*"**Florida Surgeon General** Joseph Lapado has sent a letter to the U.S. Food and Drug Administration (FDA) citing that there has been a 4,400 percent increase in reports of life-threatening conditions in Florida since the rollout of the mRNA COVID-19 vaccines."*

I've written the following to politicians and media since 20Nov2020 and updated it to 16March2023:

Total deaths increased sharply after deployment of the toxic Covid-19 "vaccines" in Jan2021 and increased further in 2022 with more vax doses. Total deaths increased in ALL age groups post-vax.

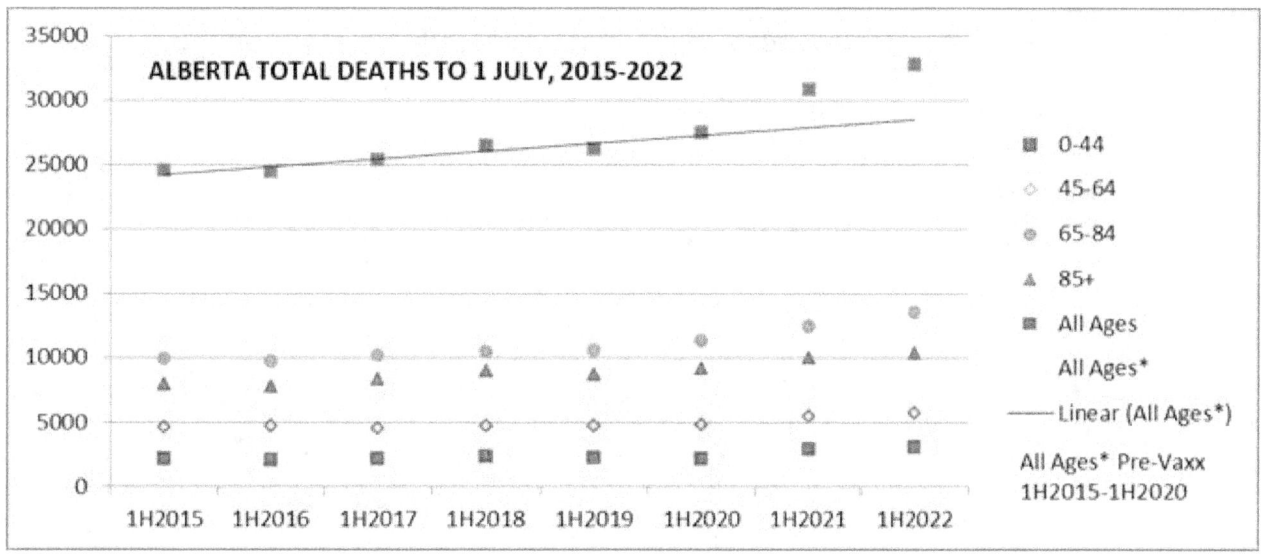

Data: Provisional weekly death counts, by age group and sex (statcan.gc.ca)

I first published this graph in 2021 – **IMPORTANT: there was no major increase in Total Deaths to well past 1Jul2020, the end of the first Covid-19 flu season**. Total Deaths are typically much greater in the Winter flu season every year.

The reason there was NO significant total death increase in Alberta in the 12 months from 1Jul2019 to 30Jun2020, the first "Covid-19 flu year", was because competent Alberta physicians practiced early treatment. Pre-Covid 2017-2018 was a worse flu-year than Covid flu-year 2019-

2020 for total deaths. **This proves that the Covid-19 illness was not a dangerous plague, was not fatal to the general population and the panic over-reaction to Covid-19 was wrong, and needlessly cost trillions of dollars and millions of lives.**

The Covid-19 "vaccines" were deployed in Alberta in Jan2021. There was a large increase in total deaths for all ages **after** the toxic Covid-19 injections were deployed.

The big increase in Total Deaths happened by 1H2021 and was caused by the toxic Covid-19 "Vaccines". There is credible evidence that this carnage was not an error, but was known to insiders in advance.

Based on total Alberta deaths:

Deaths attributed to the toxic Covid-19 "vaccines" in 2021&2022 totaled 98,000 Canadians to end 2022, increasing to 145,000 by end 2023. Justin, Jagmeet and their Liberal and NDP cohorts have now killed more Canadians than the 105,000 we lost in WW1 and WW2.

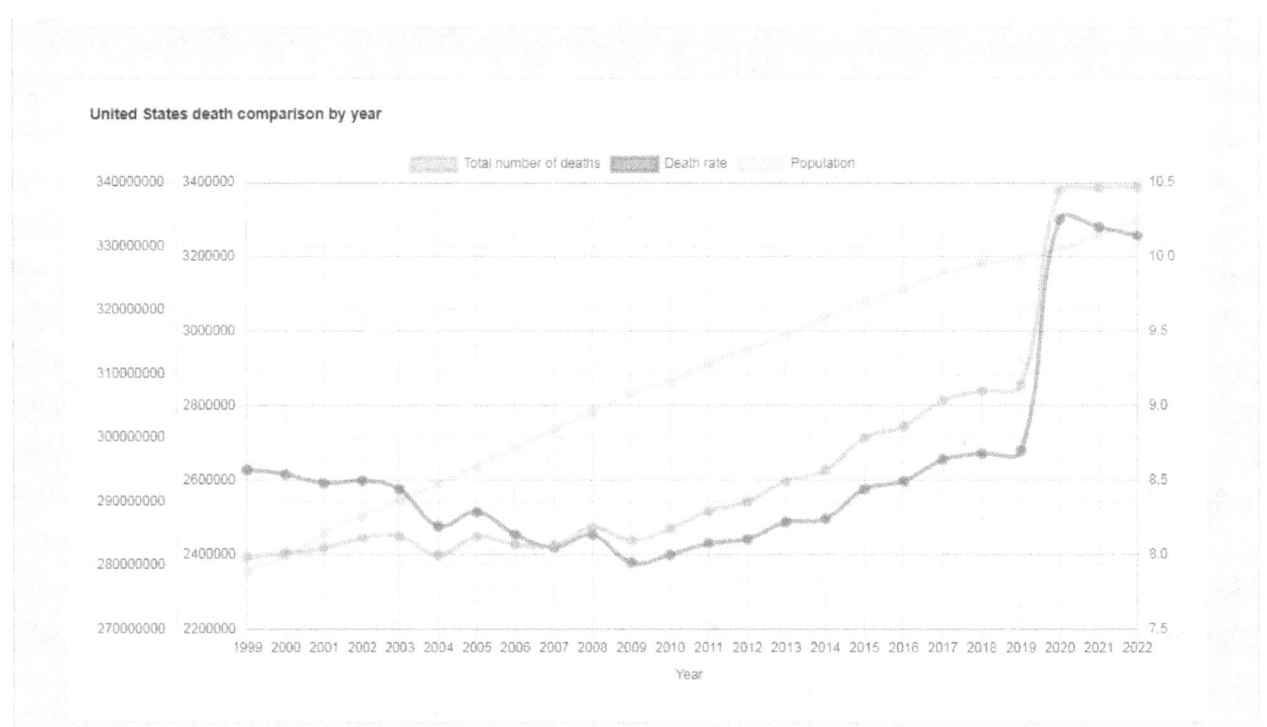

Data: <u>United States Deaths Per year, How Many Deaths in United States Per year</u>

Incompetent, late treatment of Covid-19 patients caused ~530,000 mostly-preventable American deaths. The toxic Covid-19 "vaccines" caused an additional ~1.08 million American deaths. The death toll in the USA from the mismanagement of the Covid-19 illness and the toxic Covid-19 "vaccines" totaled ~1.61 million to end 2022, and it is far from over.

Based on total USA deaths:

The death toll in the USA from the mismanagement of the Covid-19 illness and the toxic Covid-19 "vaccines" totaled 1.61 million to end 2022. The Covid-19 avoidable American death toll to date exceeds the total 1.1 million American deaths in all their wars, including the Revolutionary War, the Civil War, WW1, WW2, Korea, and Vietnam.

On 9Feb2023, **based on Alberta total deaths, I calculated 12.9 million deaths caused by the toxic Covid-19 "vaccines" in the developed countries, excluding China, India and Russia,** and this total was confirmed by Rancourt et al based on Australian and Israeli total death data. **I estimated that total vax-caused deaths in the developed countries would increase to 19 million by end 2023, for a global death toll of approximately 40 million.**

Based on current conservative Alberta rates, the 13 million total global Covid-19-vax-caused deaths in the developed countries to end 2022 will increase by ~500,000 every month unless we take action. Globally, that increase will approximate 1 million per month.

END THE HARMFUL COVID-19 LOCKDOWNS - Letters to Government and Media to 15Nov2020 (open the links for the correspondence files) End the Covid-19 Lockdowns – Harmful, Unjustified, Ineffective, as I published on 21Mar2020

Following are copies of my first series of emails to Alberta and Canadian governments and "mostly bought" mainstream media advising against the harmful, unjustified and ineffective Covid-19 Lockdowns, as I first published ~four days into the Lockdowns on 21Mar2020. None of these extreme measures were ever justified - that was obvious in Feb2020 based on quality data.

There is considerable repetition herein. There is sufficient evidence to prove that our politicians "knew or should have known" that they were committing extremely harmful, criminal acts. I initially thought the politicians were just deaf, but then realized they were "playing dumb."

Government and media were fully informed of their egregious misdeeds - they knew.

From: Allan MacRae [mailto:]
Sent: September-03-20 7:34 AM
Subject: WORLD ECONOMIC FORUM: THE GREAT RESET: A UNIQUE TWIN SUMMIT TO BEGIN 2021 (AKA "MARXISTS AT PLAY")

THE GREAT RESET: A UNIQUE TWIN SUMMIT TO BEGIN 2021

https://www.weforum.org/great-reset/about

- "The Great Reset" will be the theme of a unique twin summit in January 2021, convened by the World Economic Forum.

- "The Great Reset" is a commitment to jointly and urgently build the foundations of our economic and social system for a more fair, sustainable and resilient future.

- It requires a new social contract centred on human dignity, social justice and where societal progress does not fall behind economic development.

- The global health crisis has laid bare longstanding ruptures in our economies and societies, and created a social crisis that urgently requires decent, meaningful jobs.

- The twin summit will be both in-person and virtual, connecting key global governmental and business leaders in Davos with a global multistakeholder network in 400 cities around the world for a forward-oriented dialogue driven by the younger generation.

https://wattsupwiththat.com/2020/08/22/global-warning-the-movie-was-just-nominated-for-six-alberta-film-and-television-awards-including-best-director-documentary-feature/#comment-3067018

It's all a leftist scam - the enviro lies including the climate and green-energy fraud, the full-Gulag lockdown for Covid-19, paid-and-planned protests by Antifa and BLM. ALL LIES!

We published that the climate-and-green-energy fraud was a false narrative in 2002, and by 2012 I wrote that there was a covert agenda. Now the greens are admitting that climate-and-energy was false propaganda, a smokescreen for their extreme-left political objectives.

The green objective is to destroy prosperity and move the USA into a planned economy - with a few rich at the top looking down on the many poor peasants. That model describes most of the countries in the world. Europe and Canada are far down that path, and the USA will follow if Biden and the Demo-Marxists are elected.

The book "1984", written by George Orwell in 1949, foresaw a time "when most of the world population have become victims of omnipresent government surveillance and public manipulation".

Well here is the REAL "1984", an interview that year with ex-KGB officer and Soviet defector Yuri Bezmenov, who describes the slow, long-term "ideological subversion" of Western societies. Note Bezmenov's discussion of ideological subversion. It's about manipulating the "Useful Idiots" - the leftists in the West.

See "KGB defector Yuri Bezmenov's warning to America"

One commenter on the video wrote: "This is crazy, almost everything predicted by this guy is already happening."

COVID-19 mRNA VACCINES (PFIZER AND MODERNA) ARE HIGH-RISK – UNKNOWN FUTURE SIDE-EFFECTS - 21Nov2020 to 8Jan2021

I strongly advised our governments against deploying the toxic Covid-19 "vaccines" on 8Jan2021. 13 million killed worldwide by Covid-19 vaxes to end2022 - I was correct.

More letters to government and "mostly bought" media.
Much repetition here, but they still have not "got it"! Willfully obtuse.

I emailed the following strong warning to our governments and media on 8Jan2021.

The Covid-19 vaccine developments were rushed and are not proven safe or effective and should NOT be taken, especially by the low-risk population - those under-65 or recovered from Covid-19. The two experimental Covid-19 vaccines that contain mRNA (Pfizer and Moderna) are especially risky – due to unknown future side-effects, the risk-to-reward is far too high for the low-risk group.

Since that warning, approximately 10,000 Albertans and 100,000 Canadians have been killed by the toxic Covid-19 "vaccines". Here are the ones that hurt the most:

Fine young faces, our future, killed by the vaxes. I tried and tried to stop this…

"At a high level, corruption stems from the systemic disrespect of a nation's institutions and legal system, especially by those in power. . . .
The hallmark of virtually all corruption is 'playing dumb'."

I posted on Twitter:

> *The death toll in the USA from the mismanagement of the Covid-19 illness and the toxic Covid-19 "vaccines" totaled 1.61 million to end 2022. The Covid-19 avoidable American death toll to date exceeds the total 1.1 million American deaths in all their wars.*
>
> *Deaths attributed to the toxic Covid-19 "vaccines" in 2021&2022 totaled 98,000 Canadians to end2022, increasing to 145,000 by end2023. Justin, Jagmeet and their Liberal and NDP cohorts have killed more than twice as many Canadians as WW2 (44,000).*

I warned governments and media in March 2020! They knew or should have known.

I've watched this carnage unfold - identified it in Feb2020, published on 21Mar2020, and was ignored. 13 million Covid 19 vax-deaths in the Western world, 19 million by end 2023, ~ 40 million worldwide including China India and Russia - and it's far from over!

I'm still just trying to save lives - millions of lives.

Safeguard those you love. No more toxic Covid-19 vaxes. Treat the vaxed!

Best regards to all, Allan MacRae in Calgary

From: Allan MacRae [mailto:]
Sent: January-08-21 12:59 AM
To: health.minister@gov.ab.ca; education.minister@gov.ab.ca; 'minister.energy@gov.ab.ca'; premier@gov.ab.ca
Subject: END HARMFUL LOCKDOWNS NOW - 31 - COVID-19 mRNA VACCINES (PFIZER AND MODERNA) ARE HIGH-RISK – UNKNOWN FUTURE SIDE-EFFECTS

Hon. Shandro, Tyler
Minister of Health
health.minister@gov.ab.ca

Hon. Adriana LaGrange
Minister of Education
education.minister@gov.ab.ca

Hon. Sonya Savage
Minister of Energy
minister.energy@gov.ab.ca

Hon. Jason Kenney
Premier of Alberta
premier@gov.ab.ca

Cc: Canadian and USA media and politicians

Subject: END HARMFUL LOCKDOWNS NOW - 31 - COVID-19 mRNA VACCINES (PFIZER AND MODERNA) ARE HIGH-RISK – UNKNOWN FUTURE SIDE-EFFECTS

SUMMARY AND RECOMMENDATIONS RE COVID-19 – IMPORTANT!

- *There is no real Covid-19 pandemic. Covid-19 was only dangerous to the very elderly and infirm, and is similar in average mortality to other seasonal flu's of recent decades.*
- *The Covid-19 PCR test is not fit-for-purpose and provides many false positives. Routine testing of asymptomatic people is a waste of resources and drives erroneous policies including lockdowns.*
- *The Covid-19 lockdowns were never effective or justified. Harm done by the lockdowns exceeds by 10 to 100 times the harm from Covid-19. End all lockdowns now and do not lockdown again.*
- *Simple, inexpensive treatments are known to save lives – Vitamin D, Ivermectin etc. Why are these treatments not being widely recommended and implemented by Alberta authorities?*

- *The increase in deaths of the elderly in Winter is a well-established seasonal phenomenon. "Excess Winter Deaths" in the four Winter months routinely average about 100,000 per year in the USA and about 10,000 per year in Canada, as described on our 2015 summary of Excess Winter Mortality that includes the landmark Lancet study.*
- *NNB:*
  The Covid-19 vaccine developments were rushed and are not proven safe or effective and should NOT be taken, especially by the low-risk population - those under-65 or recovered from Covid-19. The two experimental Covid-19 vaccines that contain mRNA (Pfizer and Moderna) are especially risky – due to unknown future side-effects, the risk-to-reward is far too high for the low-risk group.

I am appalled by the policy incompetence of successive Alberta governments over the past decade and more. Fortunately for the governing UCP Party, the alternatives – NDP, Liberals and Greens would be even more destructive to Alberta. **By attempting to appease leftist extremists who seek to destroy the Alberta economy, successive Alberta governments have adopted a failed strategy that just makes us weaker and poorer.**

In the spirit of the Holidays, I wish you all a Happy New Year!

Regards, Allan MacRae, B.A.Sc.(Eng.), M.Eng.

PROVED MEDICAL FRAUD CAUSES DEATHS OF MILLIONS FROM TOXIC COVID-19 "VACCINES" / SUPPRESSION OF IVERMECTIN / DO NOT INJECT CHILDREN - 13Dec2021

The injections of children were especially unjustified – children were almost completely immune to Covid-19, until the toxic vaccines harmed their immune systems.

The Covid-19 "Lockdowns and Vaccines" strategy forced by our medical authorities and governments could not have been more wrong, resulting in ~9,000 needless Alberta deaths to end 2022, increasing to 14,000 by end 2023 unless effective treatments of the vaxed are started now. Medical murder. Willful blindness.

From: Allan MacRae [mailto:]
Sent: December-13-21 5:40 AM
To: premier@gov.ab.ca; health.minister@gov.ab.ca; education.minister@gov.ab.ca; 'CBE Office of the Chief Superintendent of Schools'; 'jason.schilling@ata.ab.ca'
Subject: PROVED MEDICAL FRAUD CAUSES DEATHS OF MILLIONS FROM TOXIC COVID-19 "VACCINES" / SUPPRESSION OF IVERMECTIN / DO NOT INJECT CHILDREN

Hon. Jason Kenney
Premier of Alberta
premier@gov.ab.ca

Hon. Jason Copping
Minister of Health
health.minister@gov.ab.ca

Hon. Adriana LaGrange
Minister of Education
education.minister@gov.ab.ca

Christopher Usih
Chief Superintendent of Schools

Calgary Board of Education
chiefsuperintendent@cbe.ab.ca

Jason C Schilling
President, Alberta Teachers Association
jason.schilling@ata.ab.ca

Cc: Canadian and USA media and politicians – **it's long past time for you to speak the truth!**

Subject: PROVED MEDICAL FRAUD CAUSES DEATHS OF MILLIONS FROM TOXIC COVID-19 "VACCINES" / SUPPRESSION OF IVERMECTIN / DO NOT INJECT CHILDREN

Covid-19 was NOT a highly dangerous global pandemic. The Covid-19 virus was only dangerous to the very elderly and infirm, as I correctly published on March 21, 2020, based on publicly available data. The world's top physicians and researchers published the same conclusion six months later in October 2020, in the <u>Great Barrington Declaration</u>.

Highly effective, inexpensive early treatments like Ivermectin were officially suppressed, resulting in millions of needless, preventable Covid-19 deaths.

By far the greatest harm to society from Covid-19 was caused by excessive government-mandated lockdowns, costly and inappropriate treatment protocols and the toxic Covid-19 "vaccines".

There is ample evidence of fraud and criminal activity in the governments' illegal over-reactions to the relatively mild Covid-19 flu.

The following important evidence supports criminal prosecutions of those responsible for the fraudulent mismanagement of Covid-19.

Regards, Allan MacRae, B.A.Sc., M.Eng.

FDA APPROVED EMERGENCY USE AUTHORIZATION FOR VACCINES DESPITE MASS CASUALTIES IN FIRST 90 DAYS

December 9, 2021

The FDA asked a federal court for a ridiculous 55-year timespan to release all of the documentation on the Pfizer COVID shots to the public. Fortunately, that judge refused and has demanded that the FDA start releasing all the info right now.

The first batch of documents have now been released and show that during the first 90 days of the vaccine rollout (December 2020 to February 2021), there were tens of thousands of adverse reactions to the shots recorded and more than 1,200 people were killed. Yet for some reason, the FDA refused to yank the shots out of circulation.

The FDA only recorded serious adverse reactions to the injections of Pfizer's vaccine, which was created in part using harvested kidney cells from aborted babies. **51,335 people suffered serious general disorders and administration site conditions; 25,957 people suffered nervous system disorders; 17,283 people suffered musculoskeletal and connective tissue disorders; 8,476 suffered skin and subcutaneous tissue disorders; 8,848 people suffered respiratory, thoracic and mediastinal disorders; 4,610 suffered from infections and – good grief – parasitic infestations (???); 5,590 people suffered from poisoning, injury and procedural complications; and 1,223 people flat-out died from being injected with the vaccine.**

The FDA knew that all this had happened in the first three months of the vaccine rollout, and had compiled all that data on adverse reactions, deaths and *parasitic infestations* by April of 2021. Just as a reminder, the experimental swine flu vaccine that was rushed to market in 1976 was yanked from the market after it killed just 25 Americans.

With the Pfizer COVID shots, the FDA knew within the first three months that the vaccine had done all this harm.

Not to get sidetracked, but what in the name of all that is holy are the parasites that are in these shots that infected 4,610 people? Parasites? Really, Pfizer?

Anyway, the FDA, which is supposed to keep the public safe from things like E coli outbreaks at the grocery store and dangerous vaccines, continued to renew the Emergency Use Authorization of these experimental shots and has now reached the point where they are about to authorize the injections for newborn babies – probably in January.

This first batch of documents from the FDA is so damning that Congressman Ralph Norman (R-SC) has introduced a bill to force the FDA to release everything publicly within 100 days. It will never pass in the House, because Nancy Pelosi really wants you to get the shot. But at least some people are starting to notice the devastating effects that these experimental shots are having on the population.

Two more professional soccer players collapsed on the field during play in Europe over the past week, by the way. Both suffered **heart attacks**. One was a 26-year-old woman and the other was a 31-year-old man. A high school football player in Montana died of a heart attack in November. And a young Equestrian rider is now hospitalized long-term with **pericarditis and peripheral neuropathy** after receiving the Pfizer jab. She has to learn to walk again.

The FDA is rushing to approve the vaccines for babies right now. Before any parent even considers having their child injected with these poisonous death shots, they should read the results of a massive study from Germany that was released just this week. Germany is the largest country in Europe, with a population of 80 million people. That includes about 10 million school-aged children under 18.

The researchers in Germany found that in the last 15 months (the duration of the pandemic), not one single healthy child in Germany has died from coronavirus. Not a single one! Only six children between the ages of 5 and 18 have died from coronavirus in Germany, and those six children all had serious, life-threatening, pre-existing conditions. All other children who contracted COVID shrugged it off, most without even showing major symptoms.

In the UK, researchers have posted similar findings. Only six children out of 12 million in the UK have died of COVID since this Chinese virus started spreading around the world. **Yet the FDA is about to tell everyone that children as young as infants can now take these dangerous, experimental shots. The German study showing that COVID poses no risk at all in children –** therefore rendering the need for children to be vaccinated moot – can be found HERE.

SMOKING GUN CONFIDENTIAL PFIZER DOCUMENT EXPOSES FDA CRIMINAL COVER-UP OF VACCINE DEATHS... THEY KNEW THE JAB WAS KILLING PEOPLE IN EARLY 2021... THREE TIMES MORE WOMEN THAN MEN

By Mike Adams

December 2, 2021

This article was originally published by Mike Adams on <u>NaturalNews.com</u>. It has been reposted here with permission from the author.

[excerpts}

Thanks to the efforts of a group called <u>Public Health and Medical Professionals for Transparency</u>, we now have smoking gun confidential documents that show Pfizer and the FDA knew in early 2021 that **Pfizer's mRNA vaccines were killing thousands of people** and causing spontaneous abortions while damaging three times more women than men.

One confidential document in particular was part of a court-ordered release of FDA files that the FDA fought by claiming the agency should have 55 years to release this information. A court judge disagreed and ordered the release of 500 documents per month, and the very first batch of documents contained this bombshell entitled, **"Cumulative Analysis of Post-Authorization Adverse Event Reports."**

Get it here:

https://phmpt.org/wp-content/uploads/2021/11/5.3.6-postmarketing-experience.pdf

....

The document reveals that within just 90 days after the EUA release of Pfizer's mRNA vaccine, the company was already aware of *voluntary* adverse reaction reports that revealed **1,223 deaths** and over **42,000 adverse reports** describing a total of **158,893 adverse reactions**. The reports originated from numerous countries, including the United States, United Kingdom, Italy, Germany, France, Portugal, Spain and other nations.

Aside from "general disorders," the No. 1 most frequently reported category of mRNA vaccine adverse reactions was **Nervous system disorders**, clocking in at **25,957** reports.

...

ALL MRNA VACCINES MUST BE IMMEDIATELY HALTED, AND FDA BUREAUCRATS MUST BE INDICTED AND ARRESTED.

IVERMECTIN CURES COVID-19 - ADULT DOSE 12mg X 5 DAYS, COST $1 PER CURE - 26Jan2022

If Ivermectin had been used as the standard treatment for Covid-10, ~13 million more people would be alive today worldwide.

Corrupt medical authorities all over the world banned Ivermectin as a treatment for Covid-19 in ~Feb 2021 in order to push the expensive, unsafe and ineffective Covid-19 "vaccines". Physicians who prescribed Ivermectin were severely punished, even though it was an effective cure for Covid-19. It can be safely assumed that these medical authorities were bribed to follow this costly, harmful path.

From: Allan MacRae [mailto:]
Sent: January-26-22 6:04 AM
To: premier@gov.ab.ca; health.minister@gov.ab.ca; education.minister@gov.ab.ca; 'CBE Office of the Chief Superintendent of Schools'; jason.schilling@ata.ab.ca
Subject: IVERMECTIN CURES COVID-19 - ADULT DOSE 12mg X 5 DAYS, COST $1 PER CURE

Hon. Jason Kenney
Premier of Alberta
premier@gov.ab.ca

Hon. Jason Copping
Minister of Health
health.minister@gov.ab.ca

Hon. Adriana LaGrange
Minister of Education
education.minister@gov.ab.ca

Christopher Usih
Chief Superintendent of Schools

Calgary Board of Education

chiefsuperintendent@cbe.ab.ca

Jason C Schilling

President, Alberta Teachers Association

jason.schilling@ata.ab.ca

Cc: Canadian and USA media and politicians

Subject: IVERMECTIN CURES COVID-19 - ADULT DOSE 12mg X 5 DAYS, COST $1 PER CURE

Many lives would have been saved and our economy much stronger if, two years ago, we had sent all these toxic Covid-19 vaccine-pushing politicians. their incompetent advisors and their "bought" mainstream media to Baffin Island - and left them there.

SCIENTIFIC COMPETENCE – THE ABILITY TO CORRECTLY PREDICT

by Allan MacRae October 20, 2021. Update November 8, 2021, Update January 14, 2022

https://correctpredictions.ca/

"The ability to correctly predict is the best objective measure of scientific and technical competence."

My predictive track record on Climate-and Covid is infinitely more accurate than the failed notions of our governments and their advisors.

For the record:

In 2002 we correctly published that there was no real global warming crisis, and grid-connected green energy schemes would fail.

My open letter to UK Undersecretary of Energy Baroness Verma in 2013 accurately predicted the current climate-and-energy crisis in Britain. The global warming-and-green-energy narrative has proved very costly, entirely false and extremely harmful.

On 21March2020 I correctly stated that the Covid-19 lockdowns were counterproductive, and we should just focus on over-protecting the very elderly and infirm – essentially the same as the Great Barrington Declaration by top experts of October 2021.

On 8January2021 I wrote our governments that the Covid-19 "vaccines" were toxic and ineffective and would cause much more harm than good. Vaccine-caused deaths and lifelong injuries have reached epidemic proportions and will continue for years. Many of the deaths and severe illnesses attributed to the Covid-19 viruses were actually caused by the toxic "vaccines".

Regards, Allan MacRae, B.A.Sc., M.Eng., Calgary

THE SCIENTIFIC MISCONDUCT STORY BEHIND IVERMECTIN

Analysis By Dr. Joseph Mercola January 26, 2022
[excerpt]

STORY AT-A-GLANCE

· In mid-February 2021, Dr. Andrew Hill at Liverpool University published a scientific meta-analysis of six randomized controlled trials involving the use of ivermectin. The review, funded by the World Health Organization and UNITAID, found the drug increased viral clearance and reduced COVID-19 deaths by 75%, yet the conclusion of the paper was dismissive

· In early April 2021, Hill was accused of scientific misconduct by the French civic group, Association BonSens. BonSens claims Hill manipulated data to downplay the usefulness of ivermectin. Hill admitted that the study sponsor had crafted the conclusion

· In early August 2021, Hill published a public notice stating one of the six studies included in his analysis had been withdrawn due to fraudulent data. A revised analysis excluding that study was published in November 2021

· In the November revision, Hill included 23 randomized clinical trials, concluding ivermectin had no statistically significant effect on survival or hospitalizations

· **Other meta-analyses of 13 to 24 studies have found reductions in death ranging from 62% to 91%. Recent research has also found a five-day course of ivermectin at a dose of 12 mg per**

day sped up viral clearance, reducing the duration of symptomatic illness by three days compared to placebo (9.7 days versus 12.7 days)

In mid-February 2021, Dr. Andrew Hill at Liverpool University published a scientific meta-analysis of six randomized controlled trials involving the use of ivermectin in 1,255 COVID-19 patients. (The paper was initially posted on a preprint server.)

The review, which was funded by the World Health Organization and UNITAID, found that ivermectin increased viral clearance and reduced COVID-19 deaths by 75%. This is a rather massive benefit, yet the conclusion of the paper was dismissive, saying additional large clinical trials were needed to make a determination about whether or not to recommend its use.

My conclusion is that IF Ivermectin or other early, competent treatments of Covid-19 had been allowed in the USA, ~530,000 people who reportedly died of Covid-19 in 2020 would still be alive. I knew in Feb2020 that Covid-19 was only fatal to the very elderly and infirm. We also knew at that time that Ivermectin was the best treatment for Covid-19 (SARS-CoV-2), based on experience from SARS CoV-1 in 2003. The FDA banned Ivermectin so it could permit the costly, ineffective, toxic Covid-19 "vaccines".

This is not just a civil law issue – this is a criminal law issue – this is medical murder for profit: Worldwide, the medical murder of ~40 million innocent people, the vax-injury of billions, and it Is far from over...

Best, Allan

APPEALS COURT RULES FDA 'EXCEEDED ITS AUTHORITY' BY ADVISING PUBLIC AGAINST USE OF IVERMECTIN TO TREAT COVID

The 5th Circuit U.S. Court of Appeals on Sept. 1 ruled the U.S. Food and Drug Administration exceeded its authority under federal law when it advised the public against using ivermectin.

By Michael Nevradakis, Ph.D.
https://childrenshealthdefense.org/defender/appeals-court-fda-ivermectin-covid/
[excerpt]

A federal appeals court last week overturned the dismissal of a lawsuit against the U.S. Food and Drug Administration (FDA), ruling that the agency exceeded its authority under federal law when it advised the public against using ivermectin.

Three doctors — Robert Apter, Mary Talley Bowden and Paul E. Marik, co-founder of the Front Line Critical Care Alliance (FLCCC) — in June 2022 sued the FDA, the U.S. Department of Health and Human Services (HHS), FDA Commissioner Robert Califf and HHS Secretary Xavier Beccera in the U.S. District Court for the Southern District of Texas.

The doctors alleged the FDA's guidance on ivermectin interfered with the doctor-patient relationship and their ability to prescribe an approved medication.

They also said their careers and professional reputations were harmed, as they faced sanctions from their employers, including suspensions and loss of privileges.

"Attempts by the FDA to influence or intervene in the doctor-patient relationship amount to interference with the practice of medicine, the regulation of which is — and always has been — reserved to states," the complaint stated.

The lawsuit also noted that the FDA approved ivermectin in 1996 for the treatment of a variety of diseases.

"The data shows the ability of the drug Ivermectin to prevent COVID-19, to keep those with early symptoms from progressing to the hyper-inflammatory phase of the disease, and even to help critically ill patients recover.

Dr. Kory testified that Ivermectin is effectively a 'miracle drug' against COVID-19 and called upon the government's medical authorities … to urgently review the latest data and then issue guidelines for physicians, nurse-practitioners, and physician assistants to prescribe Ivermectin for COVID-19[19] …

… numerous clinical studies — including peer-reviewed randomized controlled trials — showed large magnitude benefits of Ivermectin in prophylaxis, early treatment and also in late-stage disease. Taken together … dozens of clinical trials that have now emerged from around the world are substantial enough to reliably assess clinical efficacy."[20]

COVID IS THE GREATEST SCAM IN WORLD HISTORY. THIS IS A "PANDEMIC OF THE VACCINATED" – AND THE VACCINE IS KILLING US / KLAUS SCHWAB OF THE MARXIST WEF "OUTS" TRUDEAU AND CABINET - 31Jan2022

Justin and Jagmeet and their Liberal and NDP cohorts are traitors to Canada, who have caused more harm to Canadians than all our enemies in all our wars.

From: Allan MacRae [mailto:]

Sent: January-31-22 11:45 AM

To: premier@gov.ab.ca; health.minister@gov.ab.ca; education.minister@gov.ab.ca; 'CBE Office of the Chief Superintendent of Schools'; jason.schilling@ata.ab.ca

Subject: THE COVID STORYLINE IS THE GREATEST SCAM IN WORLD HISTORY. THE TRUTH IS THIS IS A "PANDEMIC OF THE VACCINATED" – AND THE VACCINE IS KILLING US / KLAUS SCHWAB OF THE MARXIST WEF "OUTS" TRUDEAU AND CABINET

Hon. Jason Kenney

Premier of Alberta

premier@gov.ab.ca

Hon. Jason Copping

Minister of Health

health.minister@gov.ab.ca

Hon. Adriana LaGrange

Minister of Education

education.minister@gov.ab.ca

Christopher Usih

Chief Superintendent of Schools

Calgary Board of Education

chiefsuperintendent@cbe.ab.ca

Jason C Schilling

President, Alberta Teachers Association

jason.schilling@ata.ab.ca

Cc: Canadian and USA ("mostly bought") media and politicians

The following article reads exactly as I warned you one year ago about toxic and ineffective Covid-19 "vaccines" - on 8 January, 2021 and since.

Ivermectin works much better than the vaccines – a quick and inexpensive cure – takes 5 days or less and costs $1 for a safe and complete at-home cure.

If Alberta had any competent leadership, they would end ALL Covid-19 mandates and TOXIC "vaccines" by tomorrow, 1 February 2022 and legalize Ivermectin.

"Many lives would have been saved and our economy much stronger if, two years ago, we had sent all these toxic Covid-19 vaccine-pushing politicians. their incompetent advisors and their "bought" mainstream media to Baffin Island - and left them there."

SCIENTIFIC COMPETENCE – THE ABILITY TO CORRECTLY PREDICT

by Allan MacRae October 20, 2021 to Present
https://correctpredictions.ca/

"The ability to correctly predict is the best objective measure of scientific and technical competence."

My predictive track record on Climate-and Covid is infinitely more accurate than the failed notions of our governments and their advisors.

Regards, Allan MacRae, B.A.Sc., M.Eng., Calgary

Subject: <u>WAYNE ROOT: THE COVID STORYLINE IS THE GREATEST SCAM IN WORLD HISTORY. THE TRUTH IS THIS IS A "PANDEMIC OF THE VACCINATED" – AND THE VACCINE IS KILLING US</u>
By Wayne Allyn Root January 29, 2022
[excerpts]

This week has been eye-opening. Even for me – and I'm the guy who has warned for over a year, in commentary after commentary, and often 3 hours a day on my nationally-syndicated radio show, that the Covid vaccine is dangerous and deadly and will lead to catastrophe.

I stuck my neck out like no other talk show host in America to warn the vaccine would not prevent illness, in fact it would damage the immune system – thereby causing more illness and death. And not just from Covid-19, but also injuries and death directly from the Covid vaccine itself.

Eight months ago, I warned it was time to suspend the vaccine program, pending an investigation of the mounting deaths, grievous injuries and permanent disabilities. I titled my commentary, "What if this experimental Covid Shot is Killing People? Don't Americans Have a Right to Know?"

Four months ago, I was courageous enough to scold New York Times medical reporters about the unfolding disaster that they have ignored. I titled my commentary, *"What I Just Told the New York Times About the Complete Failure and Disaster of the Covid Vaccine."*
...

In Alberta, Canada the government published, then quickly deleted health data exposing the fact that almost 60% of the Covid deaths classified as "unvaccinated" were actually among the vaccinated. It turns out everyone who got sick, hospitalized or died within two weeks of getting any vaccine (the first jab, or second, or third) was counted as "unvaccinated."

I'll bet you didn't know the CDC plays the exact same trick here in the USA. They know most of the Covid deaths and injuries, not to mention deaths from the vaccine itself (mostly heart attacks) will occur within 14 days of any jab. So, everyone that gets sick or dies in that period is counted as "unvaccinated."

The most perfect control group ever is the US military. Every young soldier got the Covid vaccine in the past year. To follow the results is the very definition of SCIENCE. Military whistleblowers have come forward with Department of Defense medical data showing since the start of the vaccine program cancer is up about 300% among military members; female infertility is up 500%; miscarriages are up by 300%, and there is an astronomical 1000% increase in neurological disorders from 82,000 to 863,000 in one year.

These are young men and women who were in perfect health...until the vaccines. **The vaccines are literally crippling our national defense.**

One more control group of formerly healthy young men and women- FIFA soccer players in the EU. Deaths from cardiac arrest increased by 500% in 2021. An astounding 183 professional athletes and coaches collapsed "suddenly" in 2021.

Worst of all is the news from the CDC that non-Covid deaths in the age range of 18 to 49 increased by 40% in the past year. No one has ever seen anything like this. Why are working age Americans dying in record numbers? Only one thing changed in 2021- vaccine mandates at the workplace.

The real story these numbers tell is we are experiencing a "Pandemic of the Vaccinated." The vaccine is doing catastrophic damage, but governments, politicians and bureaucrats the world over have falsely labeled the vaccinated who are sick and dying as "unvaccinated." And the media covers-up the truth like its Hunter Biden's laptop.

This is the greatest scam in world history.

IMPORTANT - <u>KLAUS SCHWAB IS SAYING HE HAS INDOCTRINATED LEADERS AND PLACED THEM IN POSITIONS (BITCHUTE.COM)</u>

In this recent Video interview with Klaus Schwab, he is bragging about all the people who have has graduated from the WEF's Young Leaders (or similar) programs who have infiltrated into positions of power. **Canada is prominent in his claims – "Over half the Trudeau Cabinet" are on board ...** Traitors! Poverty and Dictatorship!

Looks like the Kenney Cabinet is on board too. That is the only explanation for their gross mismanagement of the Covid-19 flu. No rational person or group could be this wrong for this long.

If Alberta had any competent leadership, they would end ALL Covid-19 mandates and "vaccines" by tomorrow, February 1, 2022 and legalize Ivermectin. The "vaccines" are toxic and ineffective.

I am told by reliable sources that Criminal Prosecutions are in preparation. It's time to build a prison camp on Baffin Island for these Marxist murderers and traitors.

Regards, Allan MacRae, B.A.Sc., M.Eng., Calgary

CDC DATA REVEALS VACCINES ARE KILLING FAR MORE CHILDREN THAN CHINA VIRUS ITSELF – Emails to government and media, 29Jan2022 to 11Feb2022

There was never any justification for the Covid-19 vaxing of children – who were typically asymptomatic when infected with the Covid-19 virus.

The net impact of the toxic Covid-19 vaccines on healthy people was to greatly reduce their immune system protection, causing deaths and severe illness to many. The Covid-19 vaxes were never "safe and effective" –the opposite was true.

From: Allan MacRae [mailto:]

Sent: February-11-22 5:28 AM

To: premier@gov.ab.ca; health.minister@gov.ab.ca; education.minister@gov.ab.ca; 'CBE Office of the Chief Superintendent of Schools'; jason.schilling@ata.ab.ca

Subject: CDC DATA REVEALS VACCINES ARE KILLING FAR MORE CHILDREN THAN CHINA VIRUS ITSELF

Importance: High

Hon. Jason Kenney
Premier of Alberta
premier@gov.ab.ca

Hon. Jason Copping
Minister of Health
health.minister@gov.ab.ca

Hon. Adriana LaGrange
Minister of Education
education.minister@gov.ab.ca

Christopher Usih

Chief Superintendent of Schools

Calgary Board of Education

chiefsuperintendent@cbe.ab.ca

Jason C Schilling

President, Alberta Teachers Association

jason.schilling@ata.ab.ca

Cc: Canadian and USA ("mostly bought") media and politicians

Subject: <u>CDC DATA REVEALS VACCINES ARE KILLING FAR MORE CHILDREN THAN CHINA VIRUS ITSELF</u>

I told you so, 13 months ago – Allan MacRae

Video: <u>CDC DATA REVEALS VACCINES ARE KILLING FAR MORE CHILDREN THAN CHINA VIRUS ITSELF</u>

<u>DOCTOR WHO HELPED DISCOVER OMICRON SAYS SHE WAS PRESSURED NOT TO REVEAL IT'S MILD</u>

By Jack Phillips February 10, 2022

[excerpt]

The doctor who helped discover the <u>Omicron</u> COVID-19 variant claimed that she was pressured by several government officials not to reveal that it was a milder strain.

Speaking to Germany's Welt <u>newspaper</u>, Dr. Angelique Coetzee, who is currently the head of the South African Medical Association, said that during discussions with European officials, she was told not to say that Omicron patients presented milder symptoms than prior COVID-19 variants.

"I was told not to publicly state that it was a mild illness. I have been asked to refrain from making such statements and to say that it is a serious illness. I declined," she told Welt in response to a question about her initial discussions about Omicron with European officials.

Coetzee did not elaborate on which officials allegedly told her to keep quiet. In the interview, Coetzee said that South African officials did not try to pressure her, claiming that later, she was criticized by authorities in the United Kingdom and the Netherlands.

She continued: "I am a clinician and based on the clinical picture there are no indications that we are dealing with a very serious disease. The course is mostly mild. I'm not saying you won't get sick if you're mild," according to a German-to-English translation.

"The definition of mild COVID-19 disease is clear, and it is a [World Health Organization] definition: patients can be treated at home and oxygen or hospitalization is not required," Coetzee said, adding: "A serious illness is one in which we see acute pulmonary respiratory infections: people need oxygen, maybe even artificial respiration. We saw that with Delta—but not with Omicron. So I said to people, 'I can't say it like that because it's not what we're seeing.'"

DOCTORS SMEARED BY FAUCI SPEAK OUT AGAINST LOCKDOWNS & MANDATES

I published the following in March 2020. Six months later, world experts stated the same recommendations in the Great Barrington Declaration (4Oct2021). We were all correct.

I reached this conclusion by ~1Mar2020, but only published ~3 weeks later after a discussion with one of my physician friends, who told me his 600-bed hospital was emptied to room for a "tsunami of Covid-19 patients" that NEVER ARRIVED! Nothing made sense. The Covid-19 lockdown was a SCAM from Day 1.

The difficult part of my (accurate) analysis was finding quality data in a swamp of corrupted misinformation. It was obvious from the start that Covid-19 was only very dangerous to the very elderly and infirm, and all we needed to do was over-protect them and get everyone else back to work and school. The misinformation was deliberate and agenda-driven – a world scale fraud to destroy the economies of the great democracies. The WEF, WHO, Drs Fauci & Collins and a host of corrupted government leaders and health authorities were all part of the scam.

21March2020

LET'S CONSIDER AN ALTERNATIVE APPROACH:

Isolate people over sixty-five and those with poor immune systems and return to business-as-usual for people under sixty-five.

This will allow "herd immunity" to develop much sooner and older people will thus be more protected AND THE ECONOMY WON'T CRASH.

22March2020

This full-lockdown scenario is especially hurting service sector businesses and their minimum-wage employees - young people are telling me they are "financially under the bus". The young are being destroyed to protect us over-65's. A far better solution is to get them back to work and let us oldies keep our distance, and get "herd immunity" established ASAP - in months not years. Then we will all be safe again.

I advised our Alberta and Federal governments 13 months ago on 8Jan2021 that the Covid-19 "vaccines" were TOXIC AND INEFFECTIVE and that also has proved correct. That was the second part of the Covid-19 SCAM – to peddle billions of dollars of toxic, worthless injections that have now killed or harmed more people than the Covid-19 virus. It's now time for criminal trials - Nuremberg 2.0

See my latest paper at
SCIENTIFIC COMPETENCE – THE ABILITY TO CORRECTLY PREDICT
October 20, 2021. Update February 5, 2022
https://correctpredictions.ca/
– Allan MacRae, Calgary

AUTHORS OF BARRINGTON DECLARATION SPEAK OUT
Analysis by Dr. Joseph Mercola February 11, 2022
[excerpt]

STORY AT-A-GLANCE

· October 4, 2020, three public health scientists launched The Great Barrington Declaration — a public health proposal that calls for focused protection of the most vulnerable while letting the rest of the world resume normal life

·   The Great Barrington Declaration has been signed by more than 920,000 individuals, including 46,412 medical practitioners and 15,707 scientists

·   It was recently revealed that Dr. Anthony Fauci, director of the National Institutes of Allergy and Infectious Diseases, and his former boss, now retired National Institutes of Health director Francis Collins, colluded behind the scenes to quash the declaration from Day 1

·   Focused protection is based on longstanding basic principles of public health that we have followed for decades, while lockdowns are novel, experimental strategies with no history of usefulness

·   Fauci and Collins had nothing in terms of actual science. They could not defend lockdowns or anything else based on science alone. So, they turned to propaganda, PR and smear tactics

Video: Doctors Smeared By Fauci SPEAK OUT Against Lockdowns & Mandates - YouTube

FAILED GLOBAL COVID-19 POLICIES
20Feb to 24Mar 2022

Due (without doubt) to my unrelenting pressure, Alberta Premier Jason Kenney did resign and Premier Danielle Smith took over leadership the Alberta United Conservative Party.

From: Allan MacRae [mailto:]
Sent: March-24-22 7:15 AM
To: premier@gov.ab.ca; 'minister.energy@gov.ab.ca'; health.minister@gov.ab.ca; education.minister@gov.ab.ca; 'CBE Office of the Chief Superintendent of Schools'; jason.schilling@ata.ab.ca
Subject: KENNEY MUST GO, OR THE NDP WILL WIN THE NEXT ELECTION (7) – THOSE WHO PROMOTED THE FAILED CLIMATE AND COVID-19 POLICIES WILL BE HELD ACCOUNTABLE

Hon. Jason Kenney
Premier of Alberta
premier@gov.ab.ca

Hon. Sonya Savage
Minister of Energy
minister.energy@gov.ab.ca
Hon. Jason Copping
Minister of Health
health.minister@gov.ab.ca

Hon. Adriana LaGrange
Minister of Education
education.minister@gov.ab.ca

Christopher Usih
Chief Superintendent of Schools
Calgary Board of Education
chiefsuperintendent@cbe.ab.ca

Jason C Schilling

President, Alberta Teachers Association

jason.schilling@ata.ab.ca

Cc: Canadian and USA ("mostly bought") media and politicians

Subject: KENNEY MUST GO, OR THE NDP WILL WIN THE NEXT ELECTION (7) – THOSE WHO PROMOTED THE FAILED CLIMATE AND COVID-19 POLICIES WILL BE HELD ACCOUNTABLE

Our Alberta government and the uber-corrupt NDP Opposition continue their Kabuki theatre, pretending they competently managed the Climate and Covid-19 false crises when in fact they have done enormous harm to Alberta and Albertans. This charade will not succeed, even if they control Canadian media and our justice system. Competent inquiries will be held in the USA and overseas, and their extreme mismanagement and incompetence on the Climate and Covid files will be fully revealed to all Albertans.

The great risk is that the toxic and destructive NDP will be re-elected and will continue wreak their extremist policies on Alberta, in cooperation with the traitorous Trudeau Liberals and NDP.

Jason Kenney needs to face reality, and for the good of Alberta should step aside and assist in the nomination of a competent leader for the UCP who is untarnished by the grievous errors recent years.

Regards, Allan MacRae, B.A.Sc. (Eng.), M.Eng., Calgary

'WE NEED ACCOUNTABILITY' FOR FAILED COVID-19 POLICIES: DR. SCOTT ATLAS

By Harry Lee and Jan Jekielek March 23, 2022 Updated: March 23, 2022

[excerpt]

The restrictive policies in response to the COVID-19 pandemic have largely been a failure and we need accountability for those "destructive" measures, said Dr. Scott Atlas, former special adviser to then-President Donald Trump on the coronavirus pandemic.

"We need accountability of the people who got what they wanted. They got implemented the policies they wanted. Those policies failed," said Atlas, referring to restrictive measures such as lockdowns, quarantines, school and business closures, and vaccine mandates for young people.

"We have 900,000 plus Americans who've died from COVID, according to the categorization, millions of families destroyed, yet the same people are in charge. I don't understand how that could possibly happen," Atlas told EpochTV's "American Thought Leaders" program that premiered on March 22.

According to the data from the Centers for Disease Control and Prevention (CDC), as of March 22, the COVID-19 mortality in the United States is 973,220, the highest number in the world.

"When we see that proof has come in, it's very difficult to proceed without having the accountability and the public airing to the public, to know that those strategies were wrong. It has to be admitted, and there has to be accountability or we can never restore trust in these agencies and even science itself until we get to that truth," said Atlas.

...

Atlas said when he organized a group of scientists and medical physicians to meet with Trump and then-Vice President Mike Pence, Birx pulled out at the last second. "That is not the behavior of someone who really should be at the table for looking at a complicated scientific and medical policy question. That's the behavior, in my view, of a bureaucrat," said Atlas.

It was later confirmed that Atlas invited Dr. Jay Bhattacharya, Dr. Martin Kulldorff, and Dr. Sunetra Gupta to the White House to meet Trump in Aug. 2020. The three medical experts later co-authored the Great Barrington Declaration, calling for focused protection instead of restrictive government intervention.

...

Accountability for Leaders in Federal Health Agencies and Big Pharma

Atlas said public health leaders and people in Big Pharma should also be held accountable for the failed COVID-19 response.

CONCLUSION: THE COVID-19 "VACCINES" ARE CLEARLY UNSAFE AND INEFFECTIVE!
The more times people are injected with the toxic Covid-19 "vaccines", the greater their incidence of the Covid-19 illness.

COVID "VACCINE" PERFORMANCE UPDATE | PLOTHE AND TROZZI (DRTROZZI.ORG)

In the developed world, Covid-19 "vaccine"-caused deaths totaled 13 million in 2021 and 2022, excluding China , India and Russia, as independently calculated by analysts Denis Rancourt et al and me in Feb 2023. I estimated then that total vax-caused deaths would increase to 19 million by end 2023.

The Covid-19 "vaccines" are bio-weapons, and most batches are designed for a "slow-kill". Globally, including a rough estimate for China, India and Russia, approx. 40 million people will have been killed worldwide by the toxic Covid-19 injections to end 2023, and that cull of humanity is far from over. The number of Covid-19 vax-injuries numbers in the billions.

CONCLUSION: THE COVID-19 "VACCINES" ARE CLEARLY UNSAFE!

IMPORTANT ADDENDUM:

Toxic "vaccines" for you; Safe saline for me!

Edward Dowd @DowdEdward posted:

BOMBSHELL: Pfizer employees were given a *special batch*... different from what was forced into the general population

Video: https://twitter.com/i/status/1687332318728671233

CONCLUSION: THE COVID-19 "VACCINES" ARE CLEARLY INEFFECTIVE!

This video report is about a different topic – are the Covid-19 injections even EFFECTIVE at preventing the Covid-19 illness?

COVID "VACCINE" PERFORMANCE UPDATE | PLOTHE AND TROZZI (DRTROZZI.ORG)

VIDEO: **https://drtrozzi.org/2023/08/02/covid-vaccine-performance-update-plothe-and-trozzi/**

ASIDE FROM DEATHS & INJURIES, HOW ARE C-19 "VACCINES" PERFORMING FOR COVID INFECTIONS?

MORE HARM THAN GOOD:

CONCLUSION: COVID-19 VACCINATED PEOPLE ARE BECOMING MORE ILL FROM COVID-19 THAN THOSE WHO WERE UNVACCINATED.

The First Cleveland Paper shows that the more times people are injected with the toxic Covid-19 "vaccines", the greater their incidence of the Covid-19 illness. So ask yourself why our governments and health authorities are still pushing these toxic injections?

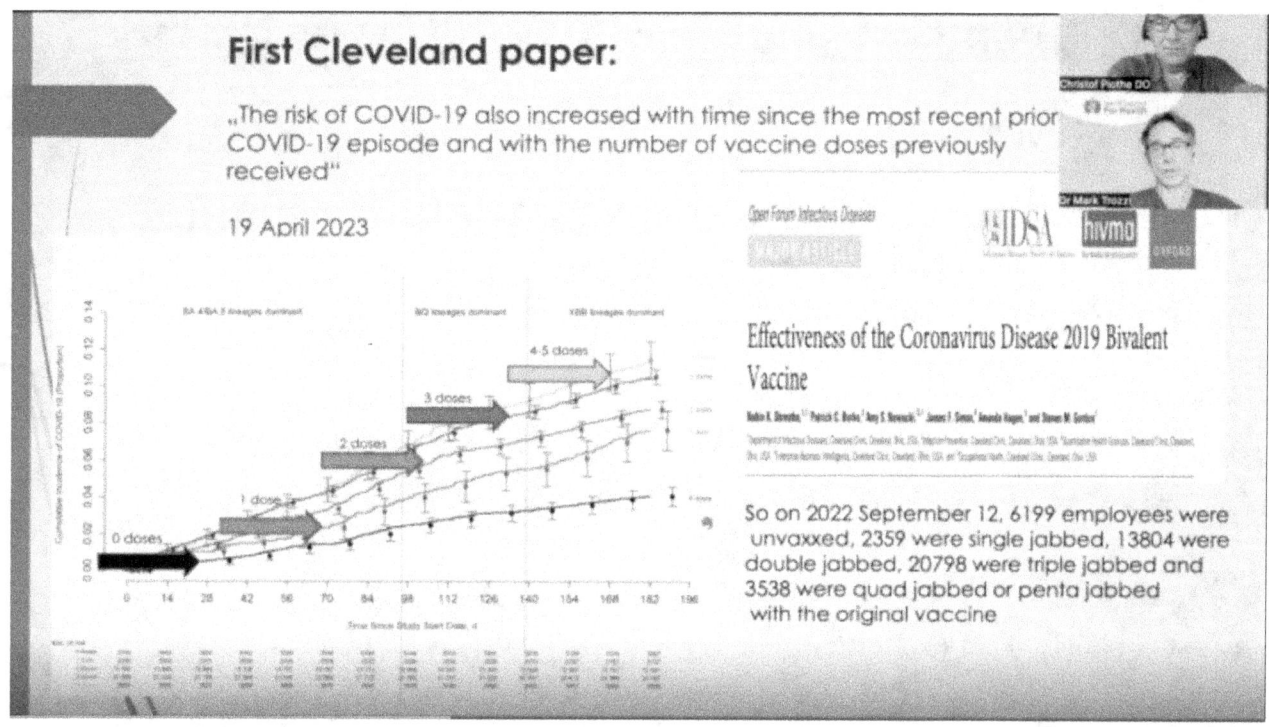

First Cleveland paper:

„The risk of COVID-19 also increased with time since the most recent prior COVID-19 episode and with the number of vaccine doses previously received"

19 April 2023

Effectiveness of the Coronavirus Disease 2019 Bivalent Vaccine

So on 2022 September 12, 6199 employees were unvaxxed, 2359 were single jabbed, 13804 were double jabbed, 20798 were triple jabbed and 3538 were quad jabbed or penta jabbed with the original vaccine

The illnesses of the Covid-19-vaxed, as well as vax-caused deaths and injuries, will accelerate this Fall as we enter flu season.

Since our corrupted politicians and health authorities are still pushing the toxic Covid-19 Injections, many gullible people will sign up for their fourth, fifth or sixth shots. Too-clever politicians and health authorities and their families will opt for the fake saline injection.

The vaxed will continue to sicken and die from the toxic "vaccines", and the health authorities will continue to falsely attribute this carnage to the virus, not their injections.

I have tried to save everyone, but there is no saving people from their own incredible naivety. Lambs to the slaughter.

I have advocated government treatment of the Covid-19-vaxed since Nov2022, but governments would still have to admit they caused this Big Cull of Humanity and that seems very unlikely. Look up The Wellness Company for treatments.

Regrets, Allan MacRae in Calgary

PHARMA IN CONTROL-ANNUAL COVID SHOTS COMING IT SEEMS

brianpeckford 1Aug2023

CDC POISED TO RECOMMEND ANNUAL COVID-19 SHOTS: DIRECTOR

By Zachary Stieber 7/30/2023
[excerpt]

The U.S. Centers for Disease Control and Prevention (CDC) is tracking toward recommending that Americans get an annual COVID-19 vaccine, the agency's new director says.

Dr. Mandy Cohen, who recently replaced Dr. Rochelle Walensky, says that the new CDC recommendation is expected to be finalized and announced in September.

"We're just on the precipice of that, so I don't want to get ahead of where our scientists are here and doing that evaluation work, but yes we anticipate that COVID will become similar to flu shots, where it is going to be you get your annual flu shot and you get your annual COVID shot," Dr. Cohen told Spectrum News.

"We're not quite there yet, but stay tuned. I think within the next couple of weeks, month we're going to hear more from our experts on COVID shots."

The director, a strong proponent of the vaccines, didn't offer any safety or efficacy data or any other details but said she worries "about parents not vaccinating kids" against COVID-19 and other viruses.

The CDC didn't respond to emailed questions, including what it would say to critics who note that there's a lack of clinical trial data supporting the shots.

Without that data, "you can't really say what the potential benefit to people is," Dr. David McCune, an oncologist, told The Epoch Times.